THAT'S MATHEMATICS

Based on lyrics by Tom Lehrer
Written by Chris Smith

Find answers, notes for parents and a glossary at the back of the book.

Illustrated by Elīna Brasliņa

ABOUT TOM LEHRER AND THIS SONG

Tom Lehrer is an American singer-songwriter, musician, satirist, mathematician and all-round genius. In 1959, when at Harvard University, he performed his "completely pointless" but utterly breathtaking song *The Elements*, featuring all the elements in the Periodic Table. Daniel Radcliffe (Harry Potter) performed this song on TV in 2020 and called Lehrer the "cleverest and funniest man of the 20th century".

Lehrer wrote around 50 songs, and summed up the connection between maths and music. "The logical mind, the precision, is the same that's involved in math as in lyrics," he said. "It's like a puzzle, to write a song."

In 2020, Lehrer granted permission for anyone to use, perform or publish his lyrics. So, as a maths-loving musician myself, I performed Tom's wonderful *That's Mathematics* song with a stellar group of enthusiastic mathematicians singing along as I played piano. It might seem weird to sing about maths but everyone in that viral video wanted to belt out the message that maths is all around us, that we use it sometimes without even realising it, and that it's a wonderful, exciting, vibrant, beautiful subject. Our song became the catalyst for this book. I hope that Tom's lyrics, Elīna's adorable illustrations and my ideas for kids and parents convince you, when you see something nifty, to announce "That's mathematics!"

Chris Smith
Scottish Teacher of the Year, 2018

See videos of Tom Lehrer and Chris Smith perform That's Mathematics *on the website www.mamamakesbooks.com*

For supporting activities, scan the QR code, or go to mamamakesbooks.com/thats-mathematics.

First published in the UK in 2023 by Mama Makes Books
ISBN 978-1-7397748-4-4 (hardback)
ISBN 978-1-7397748-5-1 (paperback)
Copyright © 2023 Mama Makes Books Ltd
Artwork and heading font © 2023 Elīna Braslina
That's Mathematics lyrics written by Tom Lehrer
Additional text written by Chris Smith
Photograph of Tom Lehrer: Associated Students, UCLA, Public domain, via Wikimedia Commons
10 9 8 7 6 5 4 3 2 1
All rights reserved, including the right of reproduction in whole or in part in any form.
A CIP catalogue record of this book is available from the British Library.
Printed in China on FSC paper.

THAT'S MATHEMATICS

Original words by Tom Lehrer
For sheet music visit tomlehrersongs.com

Counting sheep --- when you're trying to sleep,
Being fair --- when there's something to share,
Being neat --- when you're folding a sheet,
    That's mathematics!
When a ball --- bounces off of a wall,
When you cook --- from a recipe book,
When you know --- how much money you owe,
    That's mathematics!

    How much gold
    Can you hold
    In an elephant's ear?
    When it's noon
    On the moon,
    Then what time is it here?
    If you could count for a year,
    Would you get to infinity
    Or somewhere in that vicinity?

When you choose --- how much postage to use,
When you know --- what's the chance it will snow,
When you bet --- and you end up in debt,
    Oh, try as you may,
    You just can't get away
    From mathematics!
Tap your feet --- keeping time to the beat
Of a song --- while you're singing along,
Harmonize --- with the rest of the guys,
    Yes, try as you may,
    You just can't get away
    From mathematics!

THE SILLY SHEEP CIRCUS SHOW!

Decide if it's easier to count the sheep in the triangle formations or the square ones. How many sheep would you need for a bigger triangle or a bigger square?

EXPLORE THIS

One of the above groups of sheep has fallen down. Count the sheep to find out which one.

BEING FAIR WHEN THERE'S SOMETHING TO SHARE

"THAT'S NOT FAIR!"

"You've got more than me."

TRY THIS

Look at this bar of chocolate. How many pieces does it have? If three children share it fairly, how many pieces does each one get? What about two, four or six children?

*Online activity sheet available

In mathematics, sharing means splitting into equal parts. This is called **dividing**. All **even numbers** can be divided by two. So any even group of things – 6 balloons, 8 dog bones or 2 lollipops – can be shared fairly between two people.

Three each. Perfect!

I'm left with two.

EXPLORE THIS

Some numbers, like 12, are nice for sharing, but how about 5 or 7 or 11? Seven bananas can only be shared fairly with seven monkeys. Numbers like this, that can only be divided by themselves or 1, are called **prime numbers**. Can you find any more?

BEING NEAT WHEN YOU'RE FOLDING A SHEET...

... THAT'S MATHEMATICS!

TRY THIS

Get a piece of paper and fold it down the middle. What do you notice when you unfold it?

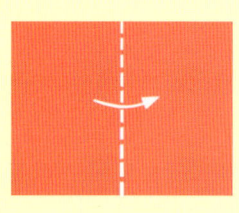

The crease splits the paper into two halves.

Fold the paper like you did before, and then fold it again.

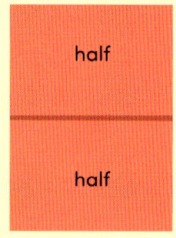

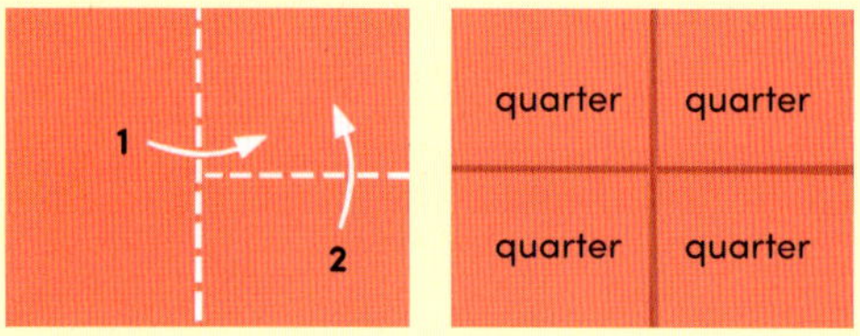

This time when you unfold it, the creases split the paper into four quarters.

What would happen if you folded it in half three times? Try it out!

Folding paper can lead to all sorts of mathematics:
- **symmetry** (do the folded sections match?)
- **fractions** (how many equal sections have we ended up with?)
- **dimensions** (what solid 3D shape do we get when we fold up this flat 2D one?)

So have fun making origami creatures or paper airplanes – it's mathematics!

EXPLORE THIS

Which flat design folds up to form each 3D shape?

*Online activity sheet available

cube   pyramid   cylinder   triangular prism

WHEN A BALL BOUNCES OFF OF A WALL

EXPLORE THIS

Look at all the **angles** written on this page. Which ones are acute, right or obtuse?

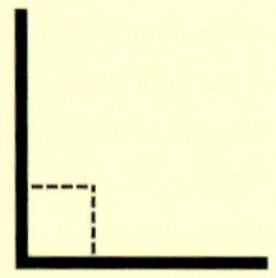

Right angles are exactly 90°.

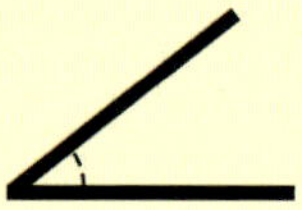

Acute angles are less than 90°.

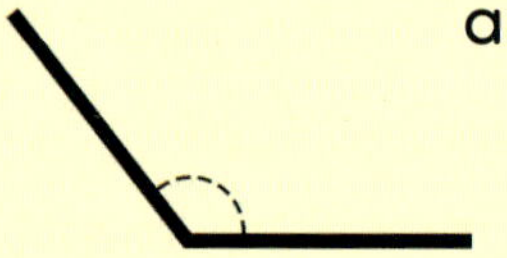

Obtuse angles are between 90° and 180°.

*Go online for a 'Right-angle Eater' who will help.

360°

40°

30° 35°
115°
60° 90° 30°

When you bounce a ball off of a wall, it rebounds at an angle. An angle is a measure of a turn. Angles are measured in degrees, using the ° symbol. When you do a complete spin, you turn 360°. A quarter turn is 90°. This is called a right angle.

1 + 1 = FUN

TRY THIS

Here is a game to put you in a spin! A caller calls out an angle and a direction to turn. Others spin to match the call. For example, "Turn 90 degrees to your right." If you get it wrong, you have to sit down. The last person standing is the winner.

Ways to make your friends dizzy:
A half spin is 180°
A whole spin is 360°
Two spins is 720°
Three spins is 1080°

WHEN YOU COOK FROM A RECIPE BOOK

TRY THIS

Which of the instuments on this page would you use to measure:
- 300ml of milk
- 500g of flour
- 15 seconds of stirring
- the temperature of jam
- the **diameter** of a round dish
- the length of a tray?

When you follow a recipe, you need to follow the ingredient list and measure the **weight, length, width, volume** or **number** of each ingredient. You have to make sure that you set the oven to the right **temperature** and your timings are spot on. There's lots of maths in cooking and baking!

EXPLORE THIS

This recipe makes enough chocolate muffins for four people. How much of each ingredient will we need if we're making double, so enough for eight people?

100g plain flour
10g cocoa powder
1 teaspoon baking powder
50g caster sugar
25g butter

1 egg
50ml milk
½ teaspoon vanilla essence
100g chocolate chunks
4 muffin cases

*Online activity sheet available

Whether you're saving up pocket money or spending it at the shops, it's important to know the value of the coins or notes you are using. You don't want to pay too much. Thankfully, it's maths to the rescue.

TRY THIS

Here's a game to try with a friend. Find twenty coins. Take it in turns to remove one, two or three coins from the pile. The person to take the final coin wins them all! Can you work out a strategy so that you always win? Do you spot any patterns?

HOW MUCH GOLD CAN YOU HOLD IN AN ELEPHANT'S EAR?

TRY THIS

Guess the length of an African elephant's ear from top to bottom. Is it about the same as your hand? Your body? The size of a tall grown-up?

When we guess a value, that's called **estimating**. You can do this in your head: it's quick and easy and you don't need to write anything down. You might not be exactly right, but the more you practise, the closer you'll get to the answer.

EXPLORE THIS

Estimate how many children will make this see-saw level. Now weigh yourself. Can you work out how many children your size would be needed?

A male African elephant can weigh 6,000kg. Here are some other things that weigh about the same.

12 grand pianos **4 cars**

WHEN IT'S NOON ON THE MOON, THEN WHAT TIME IS IT HERE?

TRY THIS

The astronaut's watch is set to Houston time, USA. Look at the three clocks opposite. What are the time differences between Houston, London, Mecca and Mumbai?

There's maths involved in telling the time to make sure you're not late for school or a party. Every minute has 60 seconds, every hour has 60 minutes, a clock face shows 12 hours and every day lasts 24 hours.

IF YOU COULD COUNT FOR A YEAR, WOULD YOU GET TO INFINITY
or somewhere in that vicinity?

TRY THIS

Set a timer. You've got 60 seconds. How high can you count in that time? How high can you count in 10 minutes?

Mmm... Really?

1, 2, 3...

I can count to infinity.

There are around 8 billion people on the planet. If you tried to count them all really quickly, you'd never get to the end, even if you lived to 100. Not only will counting to 8 billion take too long (8 billion seconds is over 250 years), but the population is growing too!

EXPLORE THIS

Let's look at some BIG numbers:

1,000	one thousand
1,000,000	one million
1,000,000,000	one billion
1,000,000,000,000	one trillion

... and these never end!
If we kept counting, we'd never reach infinity because infinity is greater than any number imaginable.

Now look at this HUGE number with 100 zeros. Can you find out what it's called (you might have to Google it)?
10,000,000,000,000,000,000,000,000,000,000,000,000,000,000,000,000,000,000,000,000,000,000,000,000,000,000,000,000,000,000,000,000,000,000,000,000,000,000,000

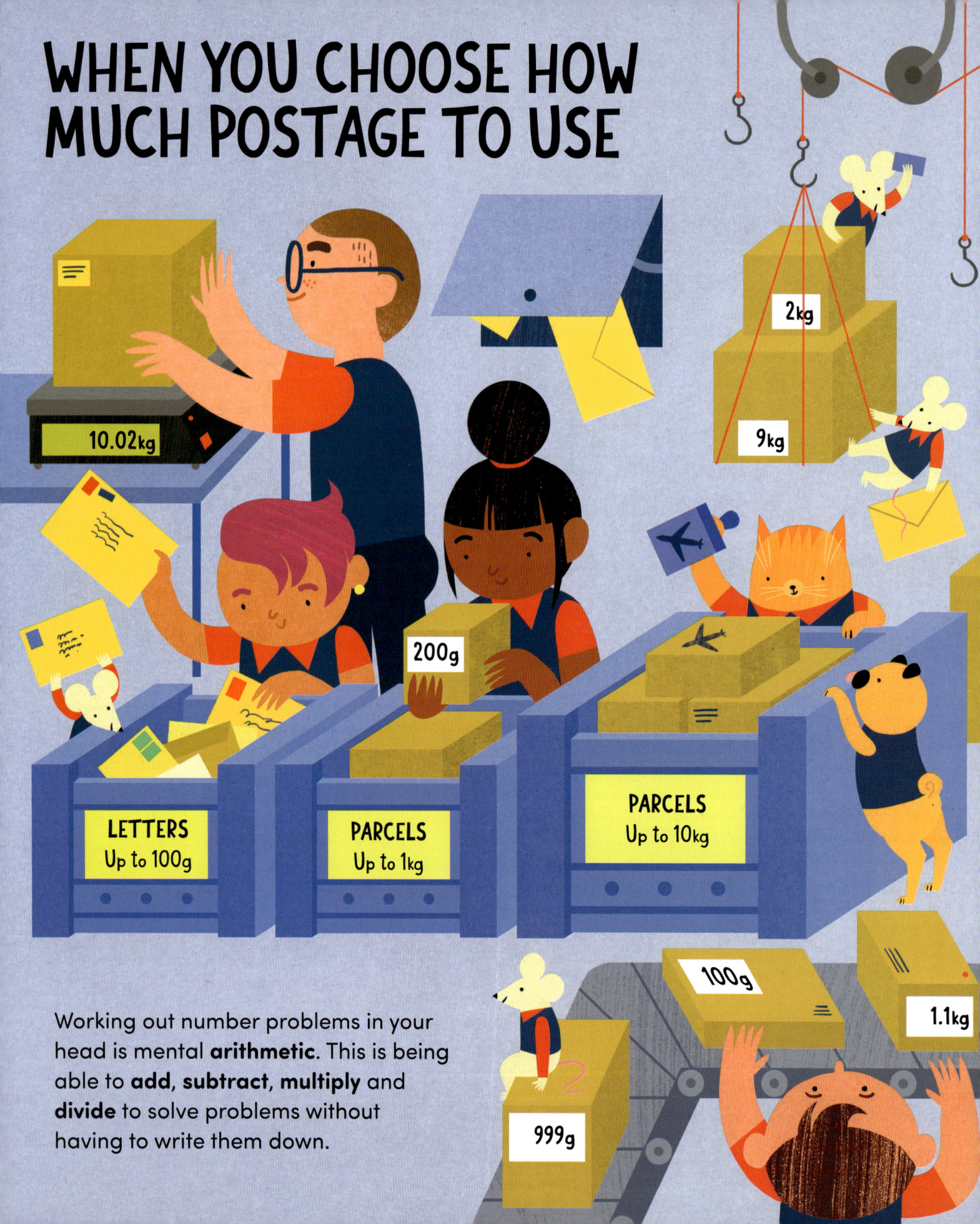

WEIGHT	POSTAGE COST
0 up to 50g	0.50
51g to 100g	1
101g to 250g	2
251g to 1kg	3
over 1kg	4
over 10kg	8

TRY THIS

Look at the parcels being sorted. Into which crates should they go and what postage should they have?

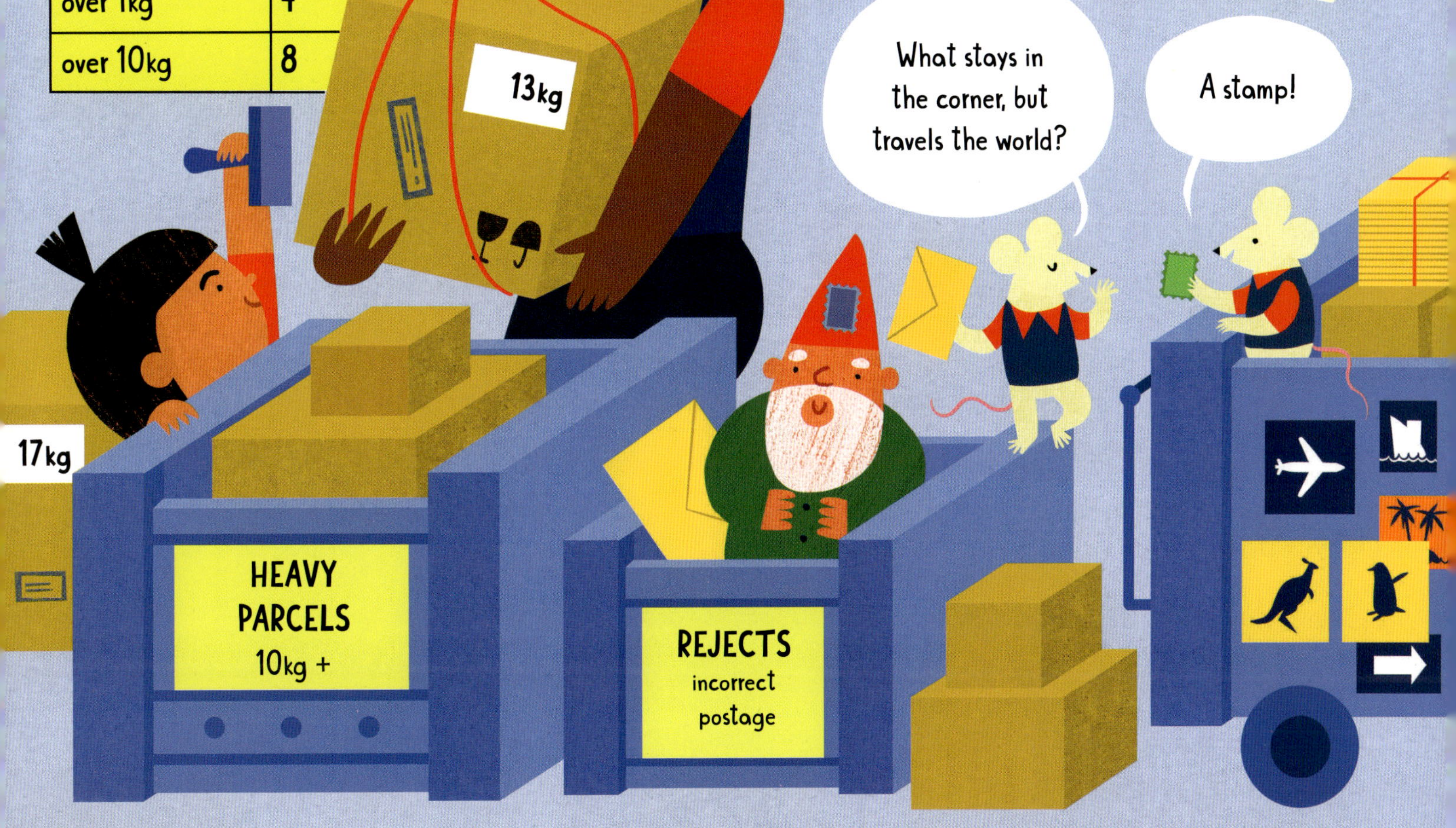

13kg
What stays in the corner, but travels the world?
A stamp!
17kg
HEAVY PARCELS 10kg +
REJECTS incorrect postage

EXPLORE THIS

1
1
2
2
4
4
8
8

Look at the value of the four stamps. Can you work out what stamps to use on each of these nine parcels, marked with the postage cost? You can only use a stamp once on any parcel.

3
5
6
7
9
10
12
13
15

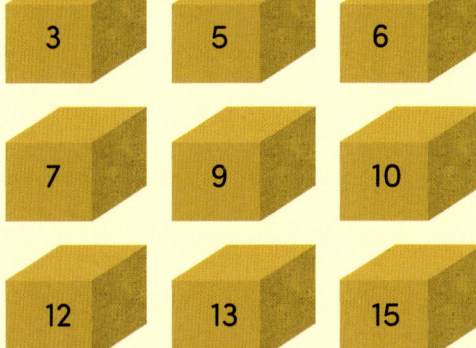

TRY THIS

Try flipping a coin, giving a spinner a whirl, or rolling dice. What are the chances of...
• flipping a head on the coin
• the spinner stopping on yellow
• rolling a five on the dice?
This is all to do with **probability**.

EXPLORE THIS

Here is the weather forecast for a week. Which day will be the warmest and which will be the coolest? Which day will be the windiest? And on which day is there the greatest chance of rain? Why not record your weather for the week?

MONDAY

28°
CHANCE OF RAIN **0%**
WIND **3 km/h**

TUESDAY

22°
CHANCE OF RAIN **5%**
WIND **6 km/h**

WEDNESDAY

23°
CHANCE OF RAIN **6%**
WIND **7 km/h**

THURSDAY

18°
CHANCE OF RAIN **20%**
WIND **14 km/h**

FRIDAY

14°
CHANCE OF RAIN **80%**
WIND **5 km/h**

SATURDAY

15°
CHANCE OF RAIN **100%**
WIND **15 km/h**

SUNDAY

9°
CHANCE OF RAIN **25%**
WIND **6 km/h**

*Online activity sheet available

TAP YOUR FEET, KEEPING TIME TO A BEAT...

To be in a band, you need to read music as well as play it, and for that you need maths. When you can count, you can keep a beat. When you know that two halves make a whole, and what a quarter is, you'll be ready to make music.

TRY THIS

Count to 20 with a steady beat, clapping your hands on **odd numbers** (1, 3, 5, 7...) and stamping your foot on even ones (2, 4, 6, 8...).

ANSWERS

How many sheep would you need for a bigger triangle or a bigger square?

15 (triangle)
25 (square)

Which group has fallen down?
There are 10 sheep so it's the biggest triangle group.

How many pieces in the chocolate bar?
12

How many pieces would each child get?
3 children = 4 pieces
2 children = 6 pieces
4 children = 3 pieces
6 children = 2 pieces

Prime numbers to 50
2, 3, 5, 7, 11, 13, 17, 19, 23, 29, 31, 37, 41, 43, 47

What would happen if you folded it three times?
You'd have 8 eighths.

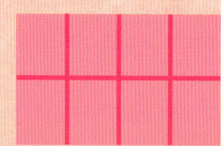

Which flat design folds up to form each 3D shape?

cube   triangular prism   pyramid   cylinder

Which angles are acute, right or obtuse?

Right angles: 90°
Acute: 30°, 35°, 40°, 50°, 60°
Obtuse: 115°
Note: Reflex angles are between 180° and 360° and 360° is one full turn.

How much of each ingredient will we need for double?

200g plain flour
20g cocoa powder
2 teaspoons of baking powder
100g caster sugar
50g butter

2 eggs
100ml milk
1 teaspoon vanilla essence
200g chocolate chunks
8 muffin cases

Which instrument would you use to measure...?

millilitres of milk = jug
grams of flour = scale or cups
seconds of stirring = timer
temperature = thermometer
length/diameter = ruler

How much money do the children need to pay for their shopping?
A = 59 (cheapest)   B = 62
C = 65 (most expensive)

Can you work out a way to win?
You will always win if you leave a multiple of 4 coins after your turn, so take enough coins to ensure you leave 12, 8 and/or 4 coins.

ANSWERS

Guess the length of an African elephant's ear.
An adult African elephant's ear is around 1.8 metres (6ft) by 1.5 metres (5ft). Measure a grown-up, arms outstretched. Are they as big?

How many children will make this see-saw level?
If you weigh 24kg, you would need 250 children like you (24 x 250 = 6,000kg). If you weigh less, you'll need more children. If you weigh more, you'll need fewer children.

What are the time differences at that moment between Houston (NASA space centre, USA), London (UK), Mecca (Saudi Arabia) and Mumbai (India)?
Houston/London = 5 hours
Houston/Mecca = 8 hours
Houston/Mumbai = 10.5 hours
London/Mecca = 3 hours
Mecca/Mumbai = 2.5 hours

Note: time differences change in summer in some (not all) countries.

Can you find out the name of the number with 100 zeros?
A googol

Note: Google was named after this number

Into which crates should they go and postage?
<u>PARCELS UP TO 1kg</u>
200g - 2, **999g** - 3, **100g** - 1
<u>PARCELS UP TO 10kg</u>
2kg - 4, **9kg** - 4, **1.1kg** - 4
<u>HEAVY PARCELS</u>
10.02kg - 8, **17kg** - 8, **13kg** - 8

Can you work out what stamps to use on each of the nine parcels?

2 + 1 = 3   8 + 2 = 10
4 + 1 = 5   8 + 4 = 12
4 + 2 = 6   8 + 4 + 1 = 13
4 + 2 + 1 = 7   8 + 4 + 2 + 1 = 15
8 + 1 = 9

What are your chances...?
Coin - 1 in 2 chances (½)
Spinner - 1 in 4 chances (¼)
Dice - 1 in 6 chances (⅙)

Which day will be...?
the warmest - Monday (28°)
the coolest - Sunday (9°)
the windiest - Saturday (15 km/h)
the greatest chance of rain - Saturday (100%)

How many beats are in each piece of music?

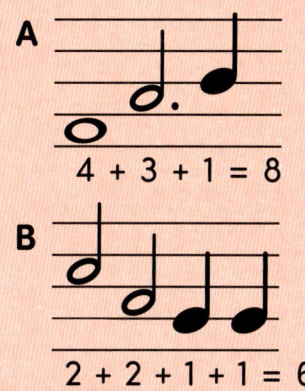

PARENTS' NOTES

This book can be enjoyed in different ways depending on a child's age and understanding. Read or sing the song and talk about the pictures, or have fun tackling the challenges together. Let children go where they want to go in the book and do what they want to do. Discussing, exploring and getting things wrong are all helpful for reasoning and problem solving.
And don't forget the free downloads on
www.mamamakesbooks.com/thats-mathematics

Counting sheep... COUNTING & NUMBER ORDER

Children learn to count by repetition, so use objects around the house to count to 10, 20, 50 and then change the value of each item. Can they count in 10s to 100 and back down again? Help them recognise the digits, and play around with order.

Being fair... FRACTIONS & DIVISION

Even young children know when a sibling gets a bigger piece of cake (a bigger fraction) or if they get less of something (divided unequally). Encourage them to share things out themselves and use words such as half, quarter, three-quarters and a whole.

Folding a sheet... SYMMETRY & SHAPES

Folding paper is a brilliant way to learn **geometry** without realising. Creating 3D shapes from flat pieces of paper is creative and satisfying but also requires care and accuracy – skills needed in maths, too.

When a ball... GEOMETRY

So many sports rely on an understanding of angles, distance, speed and rotation. Children do it with ease when they kick a ball in a goal, play copycat games or crazy golf. By drawing their attention to 'right angles' and '90 degree turns', they will pick up the foundations of geometry.

When you cook... MEASUREMENT

Most children love to cook, so let them measure out ingredients themselves. They can also use a tape measure around the house to identify things that are longer and shorter than one metre, or a timer to see how many star jumps they can do in one minute. Can they improve by practising?

When you know how much... MONEY

You can set up a play shop with labels showing the price for each item. Use real notes and coins and see if they can order them from smallest to largest in value. Using **number bonds** to 10 will help to add up in their heads.

How much gold... ESTIMATING WEIGHT & CAPACITY

Comparing things is the best way for your child to learn about weight and **capacity**. Your kitchen scales can be turned into a game of 'heavier or lighter'. Playing with cups in the bath, pans at a mud kitchen or buckets on the beach explores this concept as well. Again, use the correct maths words and they'll pick it up in no time.

When it's noon... TIME

Learning about time is complicated so introduce the idea by pointing out the positions of the hands on a clock as they go through their daily routines; play timed games; or set up races with a timer. Discuss day and night around the world.

If you could count to infinity... BIG NUMBERS

It's difficult to visualise big numbers such as a million, a billion (a thousand million) or a trillion (a million million) in real terms. One way is to compare purchasing power, e.g. on average, a car costs the same as 150 bikes and a fighter jet costs 4,000 cars. Infinity is interesting to discuss with a young child – think of the biggest number you can (a googol?) and then add one!

When you choose... CALCULATIONS & DATA

When you have a problem, often maths and/or logic will provide the answer. Developing reasoning skills helps you decide on the calculations you need to make. Give your child the time to think about how they can answer the questions on these pages.

When you know... PROBABILITY

Think of probability as how likely something is to happen. Practical activities are the best introduction. Put three red balls and one black ball in a bag. You are most likely to pick out a red ball. You have a one in four chance of picking out the black ball, which can be expressed as one quarter, ¼ or 25%.

Tap your feet... RHYTHM & MUSIC

Music is all about pattern and timing. Both listening to and playing music stimulate brain function, but playing music is proven to exercise the parts of the brain used for reasoning and complex calculations. So share music – that's mathematics, too!

GLOSSARY

add Finding the total of two or more numbers together, represented by a + symbol in calculations.

angle
   acute angle An angle greater than zero but less than 90 degrees.
   obtuse angle An angle greater than 90 degrees but less than 180 degrees.
   right angle An angle that is exactly 90 degrees.

arithmetic The maths that deals with numbers, including adding, subtracting, multiplication and division.

capacity The maximum something can contain until it is full.

diameter The straight line that joins one side of a circle to another through its centre.

dimension Flat shapes have two dimensions (we say they are 2D). They have a length and a width. Shapes you can hold have three dimensions (these are 3D). They have a length, a width and a height too.

divide To split a number or amount by another number, represented by a ÷ symbol.

estimating Using what you know to guess an amount or value.

fraction Part of a whole.

geometry The maths that deals with shapes, angles, surfaces, lines and points.

infinity A number greater than any you can imagine.

length A measurement from end to end (eg. longest side of a rectangle).

multiply When you times or multiply two or more numbers, represented by an x symbol.

number
   even number A number that can be divided exactly by 2.
   odd number A number that can't be divided exactly by 2.
   prime number A number that can only be divided by itself and 1 (1 is not considered a prime number).

number bonds to 10 Pairs of numbers that add together to make 10, such as 9 + 1 and 8 + 2.

probability The likelihood of something happening.

subtract When you take a number away from another number, respremented by a – symbol.

symmetry When you can draw or imagine a line halfway through something and one side of the line is exactly the same as the other side.

temperature How hot or cold something is, measured using degrees and represented by a ° symbol.

vicinity Near, not far away.

volume The amount of space something takes up (eg. the volume of a shape) or the quantity (eg. of a liquid).

weight How heavy something is.

width A measurement from side to side (e.g. the shortest side of a rectangle).